Inside Chocolate

Inside Chocolate

The Chocolate Lover's Guide to Boxed Chocolates

HAL AND ELLEN GREENBERG

PHOTOGRAPHS BY ALAN PORTER

Harry N. Abrams, Inc., Publishers, New York

The authors make no representations as to the size,
quality, appearance, or assortment of chocolates that you, the reader,
may encounter upon receiving any of the chocolates described herein.
The chocolates depicted in this book were supplied
by the manufacturer and may vary according
to season and availability.

EDITOR: *Anne Yarowsky*
DESIGNER: *Judith Henry*

Silk flowers by Manhattan Flower Works, New York

Library of Congress Cataloging in Publication Data

Greenberg, Hal. Inside chocolate.
Includes index.
1. Candy.   2. Chocolate.   I. Greenberg, Ellen.
II. Title.
TX791.G78   1985   664′.153   84-20528
ISBN 0-8109-1111-6 (lib. bdg.)

Published in 1985 by Harry N. Abrams, Incorporated, New York

Printed and bound in Japan

*This book is dedicated
to the house that chocolate built.*

CONTENTS

INTRODUCTION

Inside Chocolate is written for you—the dedicated, but sometimes frustrated chocolate lover. After all, who wants to bite into a seemingly delectable chocolate delight only to find inside a filling you at best dislike. To give every chocolate lover his just chocolates, we have compiled this visual directory of the variety of fillings found inside twenty-six different companies' one-pound boxes of chocolates. Because of the widespread love of chocolate, the sophistication in its manufacture, and its expense, this guide will take the worry out of choosing chocolate and help prevent the waste of even one of these delicious morsels. It is our desire that by utilizing this book, you can now pick with precision what titillates your palate and discover your very own chocolate to cherish.

Chocolate—A Brief History

Contrary to the popular belief that chocolate first originated in Europe, it is interesting to note that the Aztecs introduced it first in liquid form to the Old World, when Montezuma offered it as a gift to the Spanish explorer Cortez. The Aztecs consumed chocolate as a drink of energy and used the cocoa beans from which it was made as a source of currency.

In 1502 Columbus carried cocoa beans back with him to Spain—one of the many riches he had found in the New World. However the Spanish sought to maintain a veil of secrecy concerning chocolate's preparation until the Italians discovered its ingredients almost a hundred years later.

The mystique surrounding this brown liquid served to enhance its wide-ranging ap-

peal. Chocolate became a drink of the aristocracy and chocolate services a symbol of wealth and prestige. Giving chocolate in the interests of romance is thought to have originated in its enduring position of status and in its introduction from country to country via royal marriages. In fact chocolate first reached France through the marriage of Louis XIV to the Spanish princess Maria Theresa upon receipt of her engagement gift: a container of liquid chocolate enclosed in an ornate jewel-encrusted box. Madame Du Barry gave chocolate to her many suitors, and Casanova preferred its aphrodisiac powers to those of champagne.

In England chocolate's privileged position was maintained by the imposition of a high import duty. Furthermore, the English were the first to mix it with milk rather than imbibe it in its pure liquid form.

In the middle of the eighteenth century, Dr. James Baker brought cocoa beans back from Europe and set up his own chocolate company in America. To this day Baker's Chocolates can still be purchased in this country.

Since then chocolate has undergone many changes in use and form, one of the most significant being the creation of solid chocolate in 1876 by the enterprising Swiss Nestlés family (explaining, perhaps, why the Swiss consume more chocolate per capita than any other people).

As in history, today chocolate commands both respect and passion, status and style, romantic persuasion. On the following pages you will see and taste with the eye some of your favorite chocolates, as well as be introduced to new delights. Savor this confection in its varied flavors and forms and discriminate securely as you pursue this unique guide's taste adventure.

Inside Chocolate

1   Milk Chocolate Butter Cream
2   Dark Chocolate Coconut Cluster
3   Milk Chocolate Orange Cream
4   Dark Chocolate Vanilla Caramel
5   Milk Chocolate Peanut Straw
6   Solid Milk Chocolate
7   Dark Chocolate Vanilla Butter Cream
8   Milk Chocolate Caradip—Caramel and marshmallow
9   Dark Chocolate Marshmallow
10 Milk Chocolate Maple
11 Dark Chocolate Peanut Cluster
12 Dark Chocolate Truffle
13 Milk Chocolate Coconut Cluster
14 Dark Chocolate Raspberry Cream
15 Milk Chocolate Nougat
16 Dark Chocolate Orange Cream
17 Milk Chocolate Truffle
18 Dark Chocolate Caradip—Caramel and marshmallow
19 Milk Chocolate Strawberry Jelly
20 Dark Chocolate Raspberry Jelly
21 Dark Chocolate Mint Straw
22 Milk Chocolate Marshmallow
23 Dark Chocolate Lemon Cream
24 Almond Roca Buttercrunch
25 Milk Chocolate Vanilla Caramel
26 Dark Chocolate Nougat
27 Milk Chocolate Peanut Cluster
28 Milk Chocolate Vanilla Butter Cream
29 Dark Chocolate Butter Cream

BRiGHAM'S

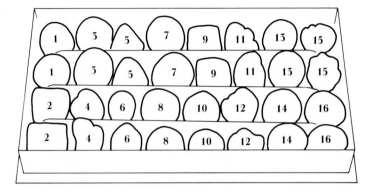

1	Dark Chocolate Cordial Cherry	9	Dark Chocolate Caramel
2	Milk Chocolate Pecan Caramel	10	Milk Chocolate Cordial Cherry
3	Rum Truffle	11	Milk Chocolate Almond Cluster
4	Dark Chocolate Cashew Cluster	12	Dark Chocolate Raisin Cup
5	Dark Chocolate Australian Apricot	13	Dark Chocolate Truffle
6	Signature Solid	14	Kahlua Truffle
7	Milk Chocolate Truffle	15	Milk Chocolate Crunchy Coconut Cup
8	Marzipan	16	Raspberry Truffle

1   Rum La Gracie—Bacardi rum flavor
     cream rolled in almonds
2   Strawberry Crème
3   Boston Butter Pecan
4   Plain Caramel
5   Honeycomb Chip
6   Cherry Cordial
7   Penoche Pecan
8   Maple Pecan
9   Rum Victoria Pecan—Brandy rum flavor
     with chopped pecans
10   Lemon Pecan
11   Vanilla Walnut
12   Peppermint Crème
13   Orange Cordial
14   Bacardi Rum
15   Black Walnut
16   Covered Nuts
17   Coconut Crème
18   Fudge Crème
19   Nut Caramel
20   Boston Butter Crème
21   Three-Flavor Neopolitan—Butter cream,
     chocolate cream, and strawberry cream
22   Undipped Caramel

C. Kay Cummings

1   Pecans
2   Brazil Nuts
3   Chocolate Nut Caramel
4   Chocolate Caramel
5   Cherry Cream
6   Chocolate Cream
7   Lemon Cream
8   Raspberry Cream
9   Vanilla Cream
10 Orange Cream
11 Apricot Cream
12 Strawberry Cream
13 Pineapple Cream
14 Cherry Cordial
15 Molasses Chip
16 Almond Chewy Nougat
17 Vanilla Chewy Nougat
18 Coconut Macaroon—Chewy coconut center
19 Swiss Mint—Chocolate mint meltaway center
20 Peanut Butter Meltaway
21 Apricot Delight—Pressed dried apricot block
22 Bordeaux Cream
23 Vanilla Caramel
24 Maple Nut Cream
25 Chocolate-Covered Almonds
26 Cashew Clusters

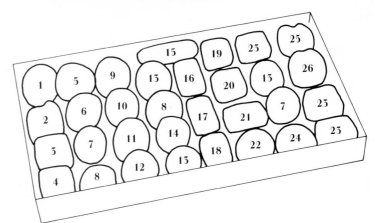

d'Orsay

1 Tosca—Milk chocolate praliné
2 Coeur Dark—Dark chocolate coffee cream
3 Ananas—Dark chocolate with pineapple-flavored chocolate cream
4 Windmill—Milk chocolate with orange peel ganache
5 Bruxelles—White chocolate with caramelized hazelnut gianduja
6 Gianduja Milk—Solid block of milk chocolate with hazelnut paste
7 Eventail—Dark chocolate lemon chocolate cream
8 Barrel—Dark chocolate with cherry mandarin, orange cream
9 Gianduja White—Solid block of white chocolate with hazelnut paste

10 Amandine—Dark chocolate almond cream
11 Désiré—Praliné swirl with glazed cherry filling
12 Ferney—Dark chocolate banana, pineapple cream
13 Rembrandt—Dark chocolate praliné and puffed rice
14 Automne—Milk chocolate with soft caramel
15 Cleopatra—Dark chocolate gianduja
16 Panier—Milk chocolate gianduja molded in paper cup and decorated with dark chocolate
17 Coeur Milk—Milk chocolate coffee cream
18 Leaf (large)—Milk chocolate chestnut praliné

19 Dessert Gianduja—White chocolate filled and topped with gianduja
20 Leaf (small)—Milk chocolate with chestnut paste
21 Chamonix—Dark chocolate hazelnut praliné dressed with praliné
22 Ruby—Dark chocolate with dark chocolate cream and cherry
23 Lady d'Orsay—Dark chocolate with wild strawberry cream
24 Versailles Milk—Milk chocolate crème caramel
25 Albert—Dark chocolate hazelnut praliné

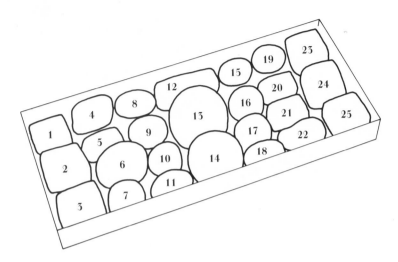

1   Mintmallow—Dark chocolate meltaway chocolate mint square topped with white marshmallow
2   Toffeemallow—Chocolate marshmallow encased in blend of ground English toffee and dark chocolate
3   Mochamallow
4   Almond Cluster—Almonds with dark chocolate
5   French Nougat
6   Turtle
7   French Cream Truffle
8   Date
9   Lemon Cream—Fresh lemon, butter, and cream

10 Swiss Truffle—Dark fudgelike chocolate center
11 Geraldine—Ground apricot and caramel
12 Pear—Half-dipped in chocolate
13 Orange—Glacéd Australian fruit half-dipped in dark chocolate
14 Apricot—Glacéd Australian fruit half-dipped in dark chocolate
15 Coffee Cream—Milk chocolate made with fresh butter and cream
16 Peanut Butter Crunchy—Peanut butter with ground peanut brittle
17 Chocolate Cream—Milk chocolate made with fresh butter and cream

18 Fudge Bar—Chocolate fudge enrobed in chocolate, topped with pecan half
19 Tosca—Ground almonds blended with honey and nutty milk chocolate
20 Molasses Chews
21 Caramel
22 Pecan Cluster
23 Caramallow—Buttery caramel topped with white marshmallow
24 Coconut Marshmallow—White marshmallow coated with blend of toasted coconut and chocolate
25 Classic Marshmallow—White marshmallow

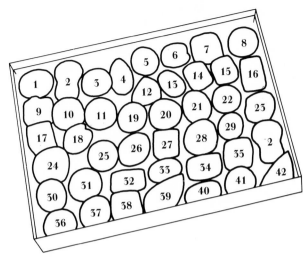

Galerie
AU CHOCOLAT™
PARIS · BRUSSELS

1 Oeuf de Mouette—Ivory chocolate gianduja
2 Champignon—Caramel and nougatine
3 Perle Noire—Nougatine and coffee-flavored ganache
4 Duja—Gianduja
5 Boule Extra—Nut praliné
6 Noix Creole—Marzipan
7 Palet Lait—Coffee-flavored ganache
8 Perle d'Or—Nougatine and ganache
9 Tabor Lait—Hazelnut praliné
10 Success—Nougatine and gianduja
11 Mogador Lait—Bittersweet chocolate cream
12 Salambo Blanc—Toasted hazelnut praliné
13 Marquise—Gianduja and whole almond
14 Bresilien—Coffee-flavored ganache
15 Favorite—Natural orange gianduja
16 Roc—Nougatine

17 Delpnine—Toffee-flavored gianduja
18 Dauphinois—Ganache
19 Crinoline Orange—Nougatine and hazelnut praliné
20 Caraque Lait—Solid milk chocolate
21 Fruit de Mer l'Ocean—Natural peach with milk chocolate center
22 Vendôme—Marzipan and nougatine
23 Fruit de Mer l'Ocean—Ivory chocolate with milk chocolate center
24 Oursin—Nougatine with hazelnut praliné
25 Palet des Neige—Cognac-flavored marzipan
26 Noix Lait—Ganache and whole walnut
27 Aiglon—Gianduja
28 Chardon Neige—Toasted hazelnut praliné
29 Blason Lait—Hazelnut and almond praliné

30 Citra—Natural orange marzipan
31 Pistra—Pistachio paste
32 Charmille—Marzipan with Grand Marnier
33 Palma—Marzipan
34 Malakoff Lait—Chopped almonds with gianduja
35 Palet Noir—Coffee-flavored ganache
36 Rocher Mignon—Praliné with chopped almonds
37 Boule Extra Lait—Nut praliné
38 Zorba Blanco—Coffee praliné
39 Fruit de Mer l'Ocean—Natural orange with milk chocolate center
40 Fondant Assorte—Vanilla cream
41 Valauris—Lemon marzipan
42 Caratendre—Butter caramel

GODIVA.
Chocolatier

1	Half Walnut—Open walnut with praliné center	**8**	Comtesse—Chocolate butter cream
2	Chestnut—Half praliné and half chocolate cream with hint of orange	**9**	Grande Mint—Dark chocolate mint cream
3	Crown—Raspberry butter cream	**10**	Chocolate Cream Leaf
4	Caramel—Dark chocolate with chocolate caramel	**11**	Leaf—Strawberry butter cream with slivers of strawberry
5	Feather—Coffee butter cream and dark chocolate sheathed in milk chocolate	**12**	Praline Almond
6	Lion of Belgium—Butterscotch caramel in chocolate	**13**	Brick—Chocolate lemon butter cream
7	Davis Cup—Orange cream in chocolate	**14**	Topaz—Coffee butter cream

15	Pyramid—Coconut cream
16	Scallop Shell—Praliné with chopped hazelnuts and brittle
17	Walnut—Butter cream, walnuts, and cherry highlights
18	Cherry Cream Square
19	Heart—Dark chocolate with chocolate cream
20	Chocolate Cream Crescent
21	Almond Butter Dome
22	Praline Swirl
23	Heart—Milk chocolate orange butter cream

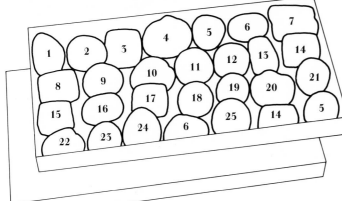

Helen Grace chocolates

1   Milk Chocolate Walnut Cluster
2   Milk Chocolate Vanilla Caramel
3   Dark Chocolate Butter Caramel
4   Milk Chocolate Almond Cluster
5   Milk Chocolate Peanut Butter Cream
6   Milk Chocolate Chocolate Butter
7   Dark and Milk Chocolate Chips
8   Dark Chocolate Pecan Delight
9   Milk Chocolate Mocha
10 Milk Chocolate Pecan Pokie—Caramel
    and pecans
11 Dark Chocolate Boysen Fiesta—
    Boysenberry cream
12 Milk Chocolate Vanilla Butter
13 Dark Chocolate French Nougat
14 Milk Chocolate Almond Nougat
15 Milk Chocolate Butter Caramel
16 Dark Chocolate Chocolate Butter
17 Milk Chocolate Almond Delight
18 Dark Chocolate Vanilla Butter
19 Dark Chocolate Lemon
20 Dark Chocolate Almond Cluster
21 Milk Chocolate Opera—Maple cream
22 Dark Chocolate Coco Almond
23 Milk Chocolate Cherry Cordial
24 Dark Chocolate Vanilla Caramel
25 Dark Chocolate Cherry Pecan

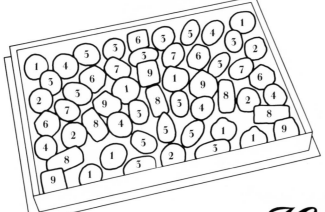

1   Hazelnut Milk Dessert
2   Dark Chocolate Crème
3   Chocolate Truffle
4   Dark Chocolate Pistachio Marzipan
5   Dark Chocolate Strawberry Crème
6   Coffee Truffle
7   Dark Chocolate Orange Crème
8   Dark Chocolate Almond Paste
9   Three-Layer Nougat

Hofbauer
VIENNA

CHOCOLATIERS SINCE 1901

1 Cherry
2 Cashew Cluster
3 Apricot Delight—Ground apricot rolled in cocoa
4 Brazil Nut
5 Nougat
6 Raspberry Jelly
7 Vanilla Caramel
8 Vanilla Cream
9 Truffle
10 Strawberry Jelly
11 Shot Ball—Rum-flavored truffle dipped in chocolate and rolled in chocolate sprinkles
12 Krunch Ball—Rum truffle dipped in chocolate and rolled in buttered nuts
13 Jordan Almond—Candy-coated almond
14 Cherry Jelly
15 Almond
16 Chocolate Caramel
17 Raspberry Cream
18 Maple Cream
19 Fancy Form—Seashell in chocolate

1 Surprise Lait—Milk chocolate liquid caramel
2 Palais Lenôtre—Dark chocolate coffee ganache
3 Gianduja—Dark chocolate with dark and milk chocolate
4 Opera—Dark chocolate caramel
5 Palais Or—Dark chocolate bittersweet ganache flavored with vanilla
6 Rocher Fondant—Dark chocolate almond and hazelnut praline
7 Rocher Lait—Milk chocolate almond and hazelnut praline
8 Charlie Fondant—Almond and hazelnut praline center covered with chopped nougatine and dark chocolate
9 Perlia—Dark chocolate caramel ganache
10 Finesse de Paris—Milk chocolate almond and hazelnut duja
11 Ecureuil—Dark chocolate hazelnut praline and chopped hazelnut
12 Petit Normand Lait—Milk chocolate with milk chocolate ganache
13 Charlie Lait—Almond and hazelnut praline center covered with chopped nougatine and milk chocolate
14 Valencia—Dark chocolate almond praline, chopped almond

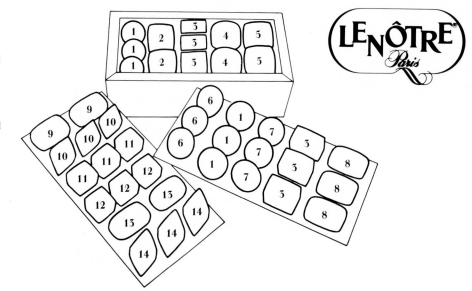

LENÔTRE Paris

Léonidas

1 Praliné Log—Milk chocolate hazelnut
and nougat praliné

2 Merveilleux Molded Dark—Milk
chocolate-flavored caramel in dark
chocolate

3 Ganache Dark—Dark chocolate orange-
flavored ganache

4 Gianduja—Solid block of milk chocolate
with chopped hazelnuts

5 Princess Milk—Milk chocolate hazelnut
praliné

6 Chestnut Praliné—Milk chocolate with
concentrated hazelnut praliné

7 Versailles Milk—Milk chocolate hazelnut
ganache

8 Pavé—Milk chocolate hazelnut praliné
and puffed rice

9 Casanova Dark—Dark chocolate hazelnut
and nougat praliné

10 Princess Dark—Dark chocolate hazelnut
praliné

11 Lingot—Milk chocolate hazelnut praliné

12 Merveilleux Molded—Milk chocolate-
flavored caramel in milk chocolate

13 Versailles Dark—Dark chocolate hazelnut
ganache

1 French Truffle	14 Raspberry Cremette
2 Chocolate-Covered Cherry	15 Fudgette
3 Mocha	16 Praline—Hazelnut crème in dark
4 French Crème	chocolate
5 Rum Rollette	17 Mocha Crème Rollette
6 Chocolate Caramel Marzipan	18 Mousse
7 Orange Fruit	19 Chocolate with Raisins
8 Coconut Dainties	20 Chocolate Dome—Solid milk chocolate
9 Bush Almond—Marzipan with truffle	21 Chocolate-Covered Pineapple
center	22 Mint Cream
10 Marzipan Acorn	23 Coffee Cremette
11 Walnut Cremette	24 French Crème Rollette
12 Caramel	25 Rum
13 Dark Chocolate with Almond	

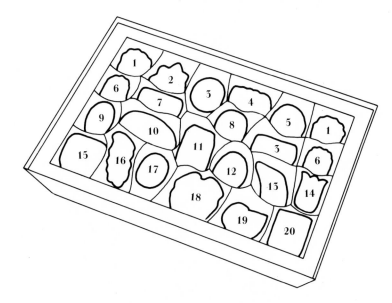

MICHEL GUÉRARD

1 Chocolate Shell
2 Solid Maple Leaf
3 Ganache
4 Nectar d'Orange—Orange in crème fraiche
5 Chocolate Crème Fraiche
6 Orange Daisy—Orange puree
7 Mousse
8 Cherry Swirl—Cherry puree
9 Coffee Crème Fraiche
10 Praliné
11 Fleur de Lis—Fruit crème fraiche
12 Cherry Cordial—Whole cherry
13 Crème Fraiche de Bruxelles
14 Coffee Tulip
15 Raspberries and Crème
16 Solid Feather Leaf
17 Caramel Noisette
18 Clover Leaf Praliné
19 Vanilla Crème Fraiche
20 Strawberries and Crème

The chocolates above are also included in the second layer.

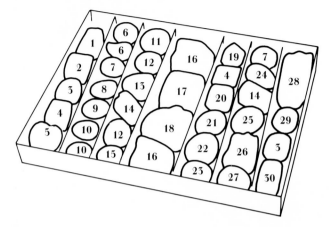

MONDEL
chocolates

1 Apricot	**16** Pecan and Caramel Patty—Almonds, walnuts, cashews, and filberts
2 Three-Layer Delight—Marzipan, bittersweet chocolate, and truffle	**17** Almond Buttercrunch
3 Gold Cup of Coffee—Solid coffee chocolate	**18** White Chocolate Nut Bark—Almonds, walnut, cashews, and filberts
4 Figaro Truffle—Ground hazelnut cream	**19** Ginger
5 Raspberry Jelly	**20** Chocolate-Covered Marzipan
6 Orange Peels	**21** Cordial Cherry
7 Mocha Truffle	**22** Grand Marnier Truffle
8 Mint Cup—Solid mint chocolate	**23** Chocolate Truffle
9 Rum—Solid chocolate	**24** Marzipan
10 Amaretto Chocolate—Almond-flavored chocolate	**25** Raspberry Cream
11 Marshmallow	**26** Raisin Bark
12 Orange Chocolate Cup—Solid orange chocolate	**27** Lemon Cream
13 Piña Colada Truffle	**28** Marshmallow Crunch—Tiny marshmallows and crispy crunch
14 Hostess Mints	**29** Coconut
15 Rum Truffle	**30** Caramel

1   Truffe Blanche—White chocolate truffle
2   Truffe Lait—Milk chocolate truffle
3   Gianduja
4   Noisette Praline—Hazelnut praline
5   Pistachio Marzipan
6   Mousse Cafe—Coffee mousse
7   Truffe Crème de Cassis—Crème de
    cassis truffle
8   Truffe Framboise—Raspberry truffle
9   Demi Noix—Walnut puree
10 Truffe Cafe—Coffee truffle
11 Gianduja Noir—Bittersweet gianduja
12 Gianduja Lait—Milk chocolate gianduja
13 Gianduja Cafe—Coffee gianduja
14 Noisette Praline Moelleux—Hazelnut
    puree
15 Truffe Noir—Bittersweet truffle
16 Truffe Champagne—Champagne truffle

MOREAU
CHOCOLATS

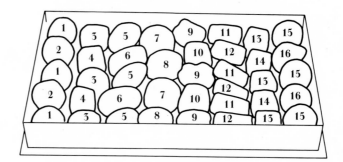

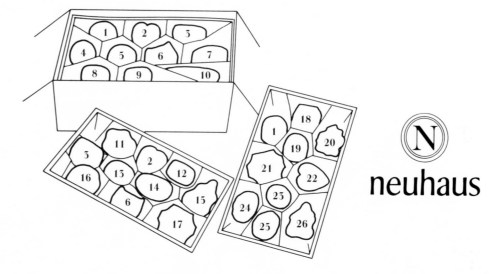

neuhaus

1 Mandarinette—Dark chocolate mandarin orange cream
2 Coeur Praline—Milk chocolate hazelnut butter cream
3 Pharaon—Milk chocolate praline with almonds
4 Cerise Kirsch—Dark chocolate cherry with kirsch
5 Les Copains—Milk chocolate praline, chopped walnut, and chestnut
6 Mephisto—Milk chocolate praline with roasted chopped walnuts
7 Sapho—Milk chocolate praline and nut paste
8 Tonneau—Milk chocolate praline gianduja with rum
9 Horse Shoe—Dark chocolate benedictine cream
10 Corne Dore—Milk chocolate gianduja in gold cornet
11 Vigne—Dark chocolate whipped egg cream
12 Tonneau—Dark chocolate sugar cream with hint of kirsch
13 Cerneau Noix—Milk chocolate praline with walnut
14 Ananas Kirsch—Dark chocolate pineapple bits with kirsch cream
15 Prestige—Milk chocolate praline and toffee
16 Horse Shoe—Milk chocolate vanilla cream

17 Le "125"—Dark chocolate pear sugar cream
18 Aurore—Dark chocolate fresh raspberry sugar cream
19 Paula—Milk chocolate praline and whole roasted walnut
20 Automne—Milk chocolate chestnut cream
21 Le "125"—Milk chocolate caramel cream
22 Coeur Grand Marnier—Dark chocolate sugar cream with hint of Grand Marnier
23 Fraise des Bois—Dark chocolate strawberry cream with bits of strawberry

24 Napoleonette—Milk chocolate praline and coffee flavor
25 Noix Double—Milk chocolate hazelnut praline
26 Apotheose—Dark chocolate butter ganache, milk, cream, and almonds

Note: For United States market, the alcohol is replaced by corresponding liquor extract in compliance with regulations set forth by the U.S. Food and Drug Administration.

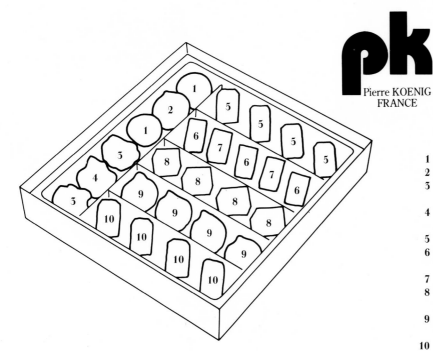

pk

Pierre KOENIG
FRANCE

1   Truffe Fondante—Bittersweet truffle
2   Truffe Lait—Milk chocolate truffle
3   Noix Pate Amande—White chocolate
    marzipan and caramel with walnut half
4   Trois Freres—Thin layer of nougatine
    with three hazelnuts
5   Blason—Milk chocolate giandujas
6   Buchette—Marzipan in bittersweet
    chocolate
7   Excellence—Hazelnut praline
8   Blanc et Noir—White and milk chocolate
    praline
9   Herisson—Giandujas covering whole
    toasted almond
10 Eucusson—Giandujas in bittersweet
    chocolate

Plumbridge INC.

1 Molasses Sponge
2 Chocolate-Covered Vanilla Caramel
3 Black Russian Mint
4 Cashew Crisp
5 Vanilla Marshmallow
6 Marzipan
7 Coconut
8 Chocolate Dipped Glacéd Australian
   Apricot
9 Crème Mint

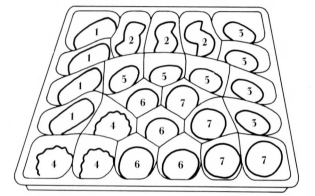

1 Belvedere—Swiss chocolate parfait
2 Lübeck—Dark chocolate marzipan
3 Windsor—Chocolate with drop of
   whiskey
4 Sevilla—Roasted split almond with milk
   chocolate
5 Wiener Cafe—Coffee liqueur
6 Truffon—Dark chocolate truffle cream
7 Florentino—Fruit gianduja

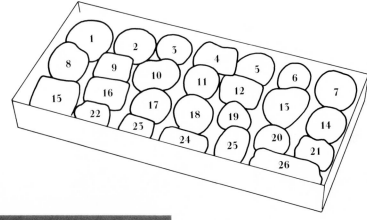

See's CANDIES

1 Milk Patties—Milk chocolate with cream, butter, and vanilla
2 Bordeaux—Milk chocolate with cream, butter, and vanilla
3 Black Walnut—Dark chocolate with black walnuts, cream, and vanilla
4 Walnut Square—Dark chocolate English walnut
5 Chocolate Butter—Unsweetened chocolate, butter, and cream in milk chocolate
6 Normandie—Bittersweet chocolate with almonds and English walnuts
7 Milk Almonds—Milk chocolate almonds
8 Dark Chocolate Almonds
9 Caramel
10 Coconut Milk—Milk chocolate with coconut, cream, butter, and vanilla
11 Date Nut—Milk chocolate date and English walnut
12 Dark Chocolate Marzipan
13 Milk Walnuts—Milk chocolate English walnuts
14 Scotchmallow—Dark chocolate marshmallow
15 California Brittle—Milk chocolate with butter and almonds
16 Dark Nougat—Dark chocolate almonds, honey, cream, and coconut
17 Vanilla Nut Cream—Milk chocolate with English walnut and vanilla

18 Bordeaux—Dark chocolate with cream, butter, and vanilla
19 Mayfair—Milk chocolate cherries and English walnuts
20 Chocolate Butter—Unsweetened chocolate, butter, and cream in dark chocolate
21 Butterscotch Square—Milk chocolate butterscotch

22 Almond Square—Milk chocolate with almonds, cream, and vanilla
23 Butterchew—Dark chocolate with cream, butter, egg whites, and vanilla
24 Rum Nougat—Milk chocolate raisins, English walnuts, and cherries with rum flavor
25 Mocha—Milk chocolate with coffee
26 Milk Chips—Milk chocolate molasses

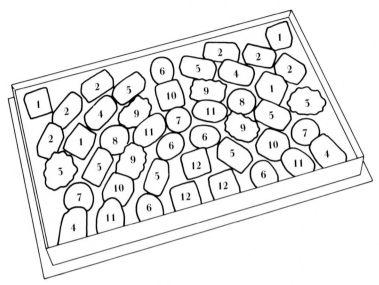

Suchard

1 Largo—Milk chocolate with soft melting filling
2 Andante—Hazelnut cream with nougat slivers
3 Agitato—Nougat praline with almonds and hazelnuts
4 Pianissimo—Dark and light gianduja with wafer crumbs
5 Vivace—Praline cream
6 Forte—Soft mocha chocolate filling
7 Presto—Hazelnut cream with whole roasted hazelnut
8 Lento—Almond cream
9 Grave—Hazelnut with almonds
10 Adagio—Fruity orange marzipan
11 Moderato—Dark hazelnut gianduja
12 Allegretto—Dark and light gianduja with hazelnut slivers

1   Troubadour—Dark gianduja
2   Surprise—Soft marzipan
3   Truffe Mocha—Soft mocha truffle
4   Tourbillon—Soft truffle
5   Merveille—Semi-dark gianduja with slivers of hazelnuts and nougat
6   Suprême—Hazelnut cream with slivers of nougat
7   Pralina—Hazelnut praline
8   Délice—Hazelnut praline
9   Fioretta—Light hazelnut cream with bitter almond flavor
10   Arlésienne—Dark gianduja with slivers of nougat
11   Florentin—Honey caramel, almonds, and hazelnuts
12   Nougatine—Nougat praline with almonds and hazelnuts

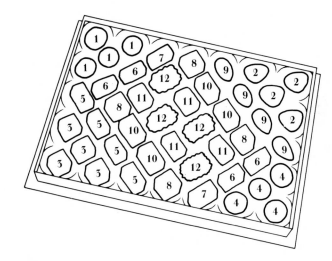

Tobler

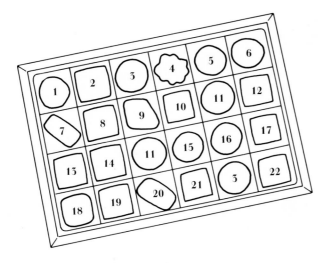

1   Marquis—Solid bittersweet chocolate
2   Milk Chocolate Square
3   Mocha Praline
4   Oakie—Clusters of nuts, Rice Crispies, and dried fruit
5   Mint Cream
6   Raspberry Jelly
7   Odile—Bittersweet truffle
8   Vanilla Square—Semi-sweet chocolate with vanilla
9   Marzipan
10   Rum Croquette—Rum truffle
11   Praline
12   Almond Crunch
13   Marinco—Truffle flavored with Cointreau, rolled in almonds

14   Orange Milk Square
15   Mint Truffle
16   Macadamia Nut
17   Aristocrat Light—Light chocolate nougat
18   Mandarin—Milk chocolate truffle flavored with orange liqueur
19   Cafe Au Lait Square—Milk chocolate with coffee
20   Hazelnut Nougat
21   Aristocrat—Dark chocolate nougat
22   Russian Caramel

This box contains approximately 3/4 lb. chocolates.

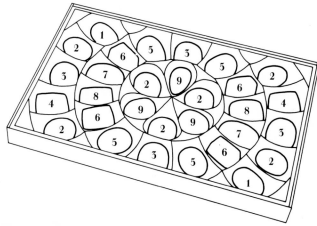

1 Berlin Mix—Fruit flavor and chocolate
2 Hazelnut Brittle
3 Nougat Praline
4 Mocha Cream
5 Crispy Caramel—Soft caramel and Rice
   Crispies
6 Soft Crunch
7 Hazelnut Cream
8 Soft Caramel
9 Mocha

VAN HOUTEN

GLOSSARY

Cream or *Sugar Cream:* Whipped sugar naturally flavored with a hint of liqueur flavor or pure fruit pulp sometimes added

Fresh Cream (Crème Fraiche): Fresh whipped cream filling

Ganache: Chocolate paste (chocolate, milk, and butter), which may be enhanced with natural flavors such as coffee, fruits, or hints of liqueur flavors

Gianduja: Finely chopped or ground nuts combined with sugar and milk chocolate

Marzipan: Confection of crushed almonds or almond paste, sugar, and egg whites

Nougat: Fruits or nuts mixed in sugar paste made either chewy or brittle

Nougatine: Nougat with chocolate covering

Praline: A sugar base to which is added a chopped nut mixture of either hazelnuts, almonds, walnuts, or a combination of these nuts

Praliné: Praline with chocolate

Truffle: Candy made of chocolate, butter, and sugar, shaped into balls and coated with cocoa, macaroon crumbs, or chopped nuts

INDEX OF RETAIL OUTLETS

Brigham's

83 "D" Plaza, The Prudential, Boston, Massachusetts 02116

342 Boylston Street, Boston, Massachusetts 02116

South Shore Plaza, Braintree, Massachusetts 02184

371 Washington Street, Brighton, Massachusetts 02135

1140 Burlington Mall, Burlington, Massachusetts 01803

15 White Street, Cambridge, Massachusetts 02140

Liberty Tree Mall, Danvers, Massachusetts 01923

Dedham Arcade, Dedham, Massachusetts 02026

37 Main Street, Hingham, Massachusetts 02043

496 Main Street, Melrose, Massachusetts 02176

New England Shopping Center, Saugus, Massachusetts 01906

91 Main Street, Stoneham, Massachusetts 02130

898 Main Street, Waltham, Massachusetts 02154

1735 Centre Street, West Roxbury, Massachusetts 02132

528 Main Street, Winchester, Massachusetts 01890

For more information contact Brigham's, Inc., 30 Mill Street, Arlington, Massachusetts 02174, (617) 648-9000.

Chocolates By M

1 Harkness Plaza, 61 West 62nd Street, New York, New York 10023, (212) 307-0777

C. Kay Cummings Candies

1987 South 1100 East, Salt Lake City, Utah 84106, (801) 487-1031

Cool's Candies

9549 Las Tunas Drive, Temple City, California 91780, (818) 286-3113

d'Orsay

Available at fine specialty and department stores coast to coast. For store nearest you call (212) 406-9270

Edelweiss Chocolates

444 North Canon Drive, Beverly Hills, California 90402, (213) 275-0341

Galerie Au Chocolat®

2420 East Camelback Road, Phoenix, Arizona 85016, (602) 954-9305

110 Dock Street, Annapolis, Maryland 21402, (301) 263-1951

The Westin Hotel, Cincinnati, Ohio 45202, (513) 421-4466

Fourth and Grant Streets, Pittsburgh, Pennsylvania 15219, (412) 391-3500

For more information call toll free 1-800-543-7679. In Ohio call (513) 381-3824

Godiva® Chocolatier

3614 The Barn Yard, Carmel, California 93920

8522 Beverly Boulevard, Suite #663, Los Angeles, California 90048

Burlington Arcade, 380 #7 South Lake Avenue, Pasadena, California 91101

Mission Valley Shopping Center, 1640 Camino Del Rio, Space #1454, San Diego, California 92108

Galleria at Crocker Center, 50 Post Street, San Francisco, California 94104

The Falls, Q-7, 888 S.W. 136th Street, Miami, Florida 33176

7429-A Dadeland Mall, North Kendall Drive, Miami, Florida 33156

Phipps Plaza, 3500 Peachtree Road, N.E., Atlanta, Georgia 30326

K301 Woodfield Mall, Schaumburg, Illinois
60195
100 Huntington Avenue, Space 15, Copley
Place, Boston, Massachusetts 02116
85 Broad Street, New York, New York 10044
560 Lexington Avenue, New York, New York
10022
701 Fifth Avenue, New York, New York 10022
Dallas Galleria, 13350 Dallas Parkway, #2305,
Dallas, Texas 75240
Capitol Hilton, 16th and K Streets, N.W.,
Washington, D.C. 20037
Georgetown Park, 3222 M Street, N.W., Wash-
ington, D.C. 20037
Washington Hilton, 1919 Connecticut Avenue,
N.W., Washington, D.C. 20009
Also available at fine department stores.

Helen Grace Chocolates

Santa Anita Fashion Park, 400 South Baldwin,
Arcadia, California 91006,
(213) 447-9901
Brea Mall, 1120 Brea Mall, Brea, California
92621, (714) 990-3022
Cerritos Mall, 251 Los Cerritos Mall, Cerritos,
California 90701, (213) 860-7211
341 Puente Hills Mall, City of Industry, Cali-
fornia 91748, (213) 965-7164
Laguna Hills Mall, 24202 Laguna Hills Mall,
Laguna, California 92653, (714) 586-1850
Bixby Knolls, 4446 Atlantic Boulevard, Long
Beach, California 90805, (213) 423-2603

Los Altos Shopping Center, 2102 Bellflower
Boulevard, Long Beach, California 90815,
(213) 596-7211
Factory Discount Store, 10690 Long Beach
Boulevard, Lynwood, California 90262,
(213) 566-8345
The City Mall, 16 City Boulevard East, Orange,
California 92668, (714) 634-0171
803 East Colorado Boulevard, Pasadena, California 91101, (213) 795-9592
Sherman Oaks Galleria, 15303 Ventura Boulevard, Sherman Oaks, California 91403,
(818) 784-1561
Del Amo Fashion Square, #69 Del Amo Fashion Square, 3526 Carson Street, Torrance,
California 90503, (213) 371-2140
Westminster Mall, Space 121, Bolsa & Ed-
wards, Westminster, California 92683,
(714) 898-1181
The Promenade Mall, 6100 Topanga Canyon
Boulevard, 252 Promenade Mall, Wood-
land Hills, California 91367, (213) 340-
1667

Hofbauer Vienna

Continental Deli, 2606 Broadway, Redwood
City, California 94063
Safeway Deli, 785 La Playa, San Francisco,
California 94121
Taste of Good Life, 705 B Wadsworth Boule-
vard, Arvada, Colorado 80005
Emily's Cheez'n & Eatables, 170 Greenwood
Avenue, Bethel, Connecticut 06801

Ridgefield Candy Shop, 440 Main Street,
Ridgefield, Connecticut 06877
Mainzer's, 12113 South Dixie Highway, Miami,
Florida 33156
Hofer's Bakery, 334 Sandy Springs Circle,
Northwest, Sandy Springs,
Georgia 30328
Continental Delicatessen, 10 South Evergreen,
Arlington Heights, Illinois 60005
Kenessey Gourmet, 403 West Belmount
Avenue, Chicago, Illinois 60657
Kuhn's Delicatessen, 116 South Waukegan
Road, Deerfield, Illinois 60648
Crossroad Trading (Gold Standard Liquor),
5100 Dempster, Skokie, Illinois 60077
Bremen House, Inc., 300 East 86th Street,
New York, New York 10028
Lips Chocolate, Greenhill Mall, Nashville,
Tennessee 37215
Neiman-Marcus, 2600 South Post Oak Road,
Houston, Texas 77056
Neiman-Marcus, 10615 Town and Country
Lane, Houston, Texas 77024
Cafe Mozart, 1331 H Street, N.W., Washington,
D.C. 20005

Krums

Emerson Shopping Center, 384 Kindkermack,
Emerson, New Jersey 07630
Riverside Square Mall, 1 Riverside Square,
Hackensack, New Jersey 08075
Clarkstown Shopping Plaza, 214 South Main
Street, New York, New York 10956

Factory Outlet, 4 Dexter Plaza, Pearl River, New York 10965, (914) 735-5100

Lenôtre Paris

Lenôtre Boutique in Sakowitz at Sakowitz Village, 5100 Beltline Road, Dallas, Texas 75240, (214) 934-8300 ext. 210

Lenôtre Shop at Northpark, 415 Northpark Center, Dallas, Texas 75225, (214) 369-4988

Lenôtre Boutique in Sakowitz at Sakowitz Post Oak, 5000 Westheimer, Houston, Texas 77057, (713) 877-8888 ext. 551

Lenôtre Shop at Town and Country, 21 Town and Country Village, Houston, Texas 77024, (713) 461-0902

Lenôtre Shop at the Production Center, 7070 Allensby Street, Houston, Texas 77022, (713) 695-9082/751-7070

Léonidas

Available at fine specialty and department stores. For store nearest you call (212) 406-9270.

Li-Lac

Barneys, 106 7th Avenue, New York, New York 10011

Chelsea Foods, 198 8th Avenue, New York, New York 10011

Li-Lac, 120 Christopher Street, New York, New York 10014, (212) 242-7374

Pleasures, 45 Pinker Street, Woodstock, New York 12498

Michel Guérard

Robinson's, 600 West 7th Street, Los Angeles, California 90017

Macy's, 170 O'Farrel, San Francisco, California 94108

Eaton's, 1 Dundas Street West, Toronto, Ontario, Canada M5B1C8

Burdines, 22 East Flagler, Miami, Florida 33101

Davison's, 180 Peachtree Street, N.W., Atlanta, Georgia 30303

Marshall Field's, 111 State Street, Chicago, Illinois 60690

Filene's, South Washington Street, Boston, Massachusetts 02101

J. L. Hudson's, 1206 Woodward Avenue, Detroit, Michigan 48226

Hahne's, 609 Broad Street, Newark, New Jersey 07101

Bloomingdale's, 160 East 60th Street, New York, New York 10022

Macy's, 151 West 34th Street, New York, New York 10001

Higbees, 100 Public Square, Cleveland, Ohio 44113

F & R Lazarus Company, High & Town Streets, Columbus, Ohio 43216

Joseph Hornes, 501 Penn Avenue, Pittsburgh, Pennsylvania 15205

Marshall Field's, 5115 Westheimer, Houston, Texas 77027

Garfinkel's, 1401 F Street, N.W., Washington, D.C. 20004

Mondel Chocolates

2913 Broadway, New York, New York 10025, (212) 864-2111

Moreau Chocolats

Encore, 3700 Old Seward Highway, Anchorage, Alaska 99503

Nordstroms, 603 D, Anchorage, Alaska 99501

C & O Steele, 7303 East Indian School Road, Scottsdale, Arizona 85251

D.D.L. Food Show, 244 North Beverly Drive, Beverly Hills, California 90210

Chocolate Affair, 55 Lafayette Circle, Lafayette, California 94549

Open Sesame, 3543 Mount Diablo Boulevard, Lafayette, California 94549

KAYLAH chocolatier, 598 Downtown Plaza, Sacramento, California 95814

Confetti, 4 Embarcadero Center, San Francisco, California 94111

I. Magnin, Geary & Stockton, San Francisco, California 94108

Fruit Basket, 1310 East 6th Avenue, Denver, Colorado 80218

Fruit Basket, 3487 South Logan, Denver, Colorado 80210

Fruit Basket, 5911 South University Boulevard, Denver, Colorado 80210

Sid's Fine Foods, 2700 South Colorado, Denver, Colorado 80222

Sweet Temptations, 164 Post Road East, Westport, Connecticut 06880

Sweet Temptations, 100 Huntington Avenue, Copley Place, Boston, Massachusetts 02116

Balducci's, 424 Sixth Avenue, New York, New York 10009

Chocolatier, Beachwood Plaza, Beachwood, Ohio 44122

Creative Chocolates, 5817 Forward Avenue, Pittsburgh, Pennsylvania 15217

Creative Chocolates, 809 Bellefonte, Pittsburgh, Pennsylvania 15232

Higgenbottom's, 3010-J West Anderson Lane, Austin, Texas 78757

Utterly Delicious, 111 Preston Royal Village, Dallas, Texas 75230

Sweet Addition, Gilman Village, Issaquah, Washington 98027

Husky's, 4721 California, S.W., Seattle, Washington 98116

— Neuhaus Chocolatier Confiseur —

Neuhaus Chocolate Shoppe, 3333 Bristol, South Coast Plaza Shopping Center, Costa Mesa, California 92626

Neuhaus Boutique at Macy's, 151 West 34th Street, New York, New York 10001

Neuhaus Chocolatier, P.O. Box 6142, Shelter Cove Harbor, Hilton Head Island, South Carolina 29938

Chocolates by Neuhaus, 13350 Dallas Parkway, Dallas, Texas 75240

Neuhaus Chocolate Shop, 427 Northpark Center, Dallas, Texas 75225

Neuhaus Chocolate Boutique, Valley View Center, Preston & LBJ Freeway, Dallas, Texas 75240

Also available at major department stores and quality stores throughout the country. For details on store locations call or write: Neuhaus (USA), Inc., 9745 Queens Boulevard, Suite 503, Rego Park, New York 11374, (212) 897-6000.

— Pierre Koenig —

Encore, 3700 Old Seward Highway, Anchorage, Alaska 99503

Nordstroms, 603 D, Anchorage, Alaska 99501

C & O Steele, 7303 East Indian School Road, Scottsdale, Arizona 85251

D.D.L. Food Show, 244 North Beverly, Beverly Hills, California 90210

Chocolate Affair, 55 Lafayette Circle, Lafayette, California 94549

Open Sesame, 3543 Mount Diablo Boulevard, Lafayette, California 94549

KAYLAH chocolatier, 598 Downtown Plaza, Sacramento, California 95814

Confetti, 4 Embarcadero Center, San Francisco, California 94111

I. Magnin, Geary & Stockton, San Francisco, California 94108

Fruit Basket, 1310 East 6th Avenue, Denver, Colorado 80218

Fruit Basket, 3487 South Logan, Denver, Colorado 80210

Fruit Basket, 5911 South University Boulevard, Denver, Colorado 80210

Sid's Fine Foods, 2700 South Colorado, Denver, Colorado 80222

Sweet Temptations, 164 Post Road East, Westport, Connecticut 06880

Sweet Temptations, 100 Huntington Avenue, Copley Place, Boston, Massachusetts 02116

Balducci's, 424 Sixth Avenue, New York, New York 10009

Chocolatier, Beachwood Plaza, Beachwood, Ohio 44122

Creative Chocolates, 5817 Forward Avenue, Pittsburgh, Pennsylvania 15217

Creative Chocolates, 809 Bellefonte, Pittsburgh, Pennsylvania 15232

Higgenbottom's, 3010-J West Anderson Lane, Austin, Texas 78757

Utterly Delicious, 111 Preston Royal Village, Dallas, Texas 75230

Sweet Addition, Gilman Village, Issaquah, Washington 98027

Husky's, 4721 California, S.W., Seattle, Washington 98116

Plumbridge, Inc.

30 East 67th Street, New York, New York 10021, (212) 371-0608

Sarotti®

Beverly Wilshire Hotel, 9500 Wilshire Boulevard, Beverly Hills, California 90212

Ernie's Continental Delicatessen and Imports, 8400 8th Avenue, Inglewood, California 90305

Le Drug Store, Marina Center, 4718 Admiralty Way, Marina del Rey, California 90291

Henshey's Department Store, 402 Santa Monica Boulevard, Santa Monica, California 90401

Potpourri, 2305 Wilshire Boulevard, Santa Monica, California 90403

Eschbachs European Delicacies, Alpine Village, 833 West Torrance, Torrance, California 90502

Robert's of Woodside, 3015 Woodside Road, Woodside, California 94062

Gelson's Gourmet Market, Southern California locations

Bremen House, 300 East 86th Street, New York, New York 10028

Brite Glo Products, 66 Nagle Avenue, New York, New York 10040

Food Emporium, all East Coast locations

Grand Union Supermarkets, all locations in New York, New Jersey, and Connecticut

Duty Free Shops at airports in New York, Los Angeles, and San Francisco

For other locations near you call (213) 935-6667

See's Candies

Shops listed in major cities in the following states: Arizona; California; Colorado; Hawaii; Illinois; Missouri; Nevada; New Mexico; Oregon; Tennessee; Texas; Utah; Washington.

Check your local white or yellow pages for shop nearest you.

Suchard

For shop nearest you, write: Tobler/Suchard USA, 1400 East Wisconsin Street, Delavan, Wisconsin 53115, or call: Tobler/Suchard, (201) 227-8070

Tobler

For shop nearest you, write: Tobler/Suchard USA, 1400 East Wisconsin Street, Delavan, Wisconsin 53115, or call: Tobler/Suchard, (201) 227-8070

Valentine's Cosmopolitan Confections

1112 4th Street, San Rafael, California 94901

Van Houten

Fedco Stores:

11525 East South Street, Cerritos, California 90701

3535 South La Cienega Boulevard, Los Angeles, California 90016

3111 East Colorado Boulevard, Pasadena, California 91107

14920 Raymer, Van Nuys, California 91405

Eleven States Trading, 852 North La Brea, Hollywood, California 90038

Long's Drug Store, 12201 Victory Boulevard, North Hollywood, California 91606

Monel, Inc., 4238 North West 37th Avenue, Miami, Florida 33142

Chicago Importing, 1144 West Randolph Street, Chicago, Illinois 60607

Snider Drug, 905 Yankee Doodle Road, Eagan, Minnesota 55121

Mel Markets, 594 Atlantic Avenue, East Rockaway, New York 11518

Ogontz Sales, 2201 East Cambria Street, Philadelphia, Pennsylvania 19125